OSCAR THE SCARED OCTOPUS

Written and Illustrated by: Terrie L. Birmingham

This book is dedicated to my family and friends and in loving memory of my grandpa, Oscar (even though he wasn't a fan of the water).

OSCAR THE SCARED OCTOPUS

Written and Illustrated by:

Terrie L. Birmingham

Oscar was in his cave, patiently waiting for his afternoon visit with his friend Clara Clownfish.

Every day since they met, Clara has made a special point to come visit him because she didn't want him

to be lonely. "Oscar, I was invited to go to Sammy Stingray's surprise birthday party. You are invited

too. Would you come with me? It would be a lot of fun" Clara said.

"No thank you. You know I don't like to leave the safety of my cave. What you don't know is that I get scared. I was afraid to tell you that I am scared. I thought maybe you wouldn't want to be my friend anymore if you knew that I get scared sometimes" said Oscar.

"What are you so afraid of? How will you ever make any other friends or have any fun if you never leave your cave?" Clara asked.

Oscar told Clara all about predators such as sharks, eels and dolphins that are in the ocean. He said "If we leave the safety of my cave to go into open waters we might be an easy target for them to find. I would rather stay here, inside the cave, so we will be safe."

"Okay Oscar, I won't ask you to leave your cave anymore. I can understand why you might be scared and would prefer to stay here in the cave where it is safe. I will tell you a secret, if you promise not to laugh at me," said Clara. "Even I get scared sometimes. If I see an eel I usually scream really loud. They are the creepiest thing I can think of! Oscar, I want you to know something. You never have to be afraid to tell me something. I am your friend. I would never laugh at you. Would you like to hear something funny?"

"Sure I would," he said.

"I have always wished I could be an octopus" she told him.

He asked her what she knew about his species.

"You are the only octopus that I know. I'm sure there is a great deal about your species that I don't know. Would you tell me about Octopuses?"

"Sure I would love to tell you about my species. The first thing I can say is that when there is more than one octopus you say Octopi" replied Oscar. "As you already know our natural predators are sharks, eels and dolphins. We are large compared to other species but that certainly doesn't make us invincible. We do have some built in defense mechanisms, although they don't always save us from danger. Some of our defense mechanisms help us escape from predators, if we are ever in a dangerous situation." Clara asked, "What defense mechanisms do you have Oscar?"

"If a shark or other predator grabs a hold of my arm, I can detach it and escape. My arm will grow back eventually."

Clara started to laugh. "Oscar, you silly thing, you don't have arms, you have tentacles." "Actually Clara, you are incorrect. Tentacles are longer than arms and a tentacle only has only one sucker at the end of it. We Octopi have arms that are covered in suckers. They are used to grab as well as taste and feel."

"Our arms and body are made up of skin cells called chromatophores that help us to blend into the background, like being camouflaged. Octopi are kind of like chameleons. We are able to change our skin color and our skin texture to blend into our surroundings, so that our enemies can't see us" Oscar told Clara.

"Last night I was hungry. I went out, into the open water, in search of something to eat; like shell fish or some mollusks. I saw something shiny off in the distance. I decided to take a closer look to see what was shining. As I approached, I suddenly realized I was looking at a sunken treasure chest."

"Oh that's exciting. What kind of treasure was in it?" she asked.

Oscar said, "I don't know. I noticed an eel lurking around the base of the treasure chest."

"What did you do?" Clara asked.

Oscar told her he slowly swam backwards. "I spotted some yellow coral with red dots nearby. I made my skin change to match the coral. It worked perfectly. The eel never saw me. I swam away before it spotted me."

"On my way back home, to my cave, I did run into another close encounter. Just as I swam around the coral reef, near my home, I saw two sharks swimming. They noticed me almost immediately." "Oh no, what happened?" gasped Clara.

"Don't worry, Clara, I escaped unharmed."

"How did you do that?" she asked.

"I inked them," said Oscar. Clara asked Oscar what he meant by inking them. "We octopi have

the ability to squirt out ink to defend ourselves," he said.

"How do you do that?"

"It's simple really, most species of octopi have the ability to release a thick cloud of black ink that helps to confuse the predators and provide the octopus an edge when they are trying to escape. When I release the ink cloud, the predator can no longer see me. The ink also stops them from smelling me. I can swim away before the cloud disappears."

"Wow Oscar, it must be pretty cool to be an octopus." Clara said.

"You are correct; some things are pretty cool," he said. "Did you know that octopi are thought to be the smartest of all the invertebrates? They are also known for having good problem solving skills."

"One thing you probably did not know I that we can fit into small crevices, because we have no bones or internal shells. We squeeze ourselves into those tiny spaces."

"Another thing that is pretty neat, and way different from most species, is the fact that we have three hearts instead of just one. Our blood color is not red; it is a shade of blue."

Then Clara says, "I wish I could be an octopus, instead of a clownfish. Clownfish can't do anything cool like re-growing arms, squirting ink or having the ability to camouflage themselves."

"Not everything is so cool for us Octopi. Did you know that we have short lifespan? Our lifespan is between six months and five years."

"Why is that Oscar? Why don't most octopi live longer than five years?" Clara asked.

"When female octopi lay their eggs it usually takes about a month for them to mature and hatch.

The females guard their offspring for the entire time. They never leave the eggs until they are hatched.

They do not even hunt during the time that they are guarding their eggs. It is very common for them to die from starvation. Those who do not die become very weak and can't escape from predators.

Clara gasped, "Wow, I did not know that. I am glad that you told me all about your species. Now, will you please reconsider coming to the birthday party with me? I really don't want to go alone!"

"I suppose I will go with you so you don't have to go alone" replied Oscar.

"It's almost time for the party to start. How long will it take you to get ready?" I am ready to go now, but I don't have a gift for Sammy" he said.

"I brought two gifts for Sammy. You can give her one of them."

"Thank you Clara! I am so lucky to have a friend like you." "I am glad to have you as a friend too. Thank you so much for coming to the party with me. I am sure you will have a good time," said Clara. "Sammy's house is not too far from here. We should leave in a few minutes so that we are not late for the party."

"Oscar, Sammy is going to be so surprised to see you! Everyone will be surprised to see you. It's not often you leave your cave. I think this might just turn out to be the best surprise party for everyone!" Oscar laughed and said "You are probably right.

The End.

About the Author/ Illustrator:

Terrie L. Birmingham

Terrie Birmingham has been teaching art for the Greater Amsterdam School District since 2001. She holds a Masters of Science in Teaching and Learning Degree, a Bachelors of Science in Art Education and an Associates Degree in Art.

Terrie owns and operates a tattoo parlor (Tattoo Artist at Large).

When Terrie is not teaching or giving tattoos she likes to paint, draw, make quilts, create murals and a wide variety of other artistic things. Some of her work can be seen at the college of St. Rose in their permanent art collection.

Terrie lives in upstate New York with her two children (BJ and Brittany) husband (Kevin) and three dogs.

www.ingramcontent.com/pod-product-compliance
Lightning Source LLC
Chambersburg PA
CBHW050432180526
45159CB00006B/2505